Forewords

"¹¹ God said, "Let the earth yield grass, herbs yielding seeds, and fruit trees bearing fruit after their kind, with their seeds in it, on the earth;" and it was so. ¹² The earth yielded grass, herbs yielding seed after their kind, and trees bearing fruit, with their seeds in it, after their kind; and God saw that it was good. ¹³ There was evening and there was morning, a third day."
(World Messianic Bible, Genesis)

Life is really mysterious for me. We found such a various types of living things everywhere on the earth, and lots of species of living organisms perform their life works everyday everywhere in the sea, on the land, and in the sky. It is definitely impossible to know and tell about everything of the living things on earth. As a research scientist of life science for more than thirty years, I could not stop feeling awe on the wonderfulness and beautifulness of living things.

We still do not know anything on the life in space outside the earth. While human beings have developed such a great technology to investigate planets, we have not yet identified any sign of living things on planets other than the earth. Life borne and grown on earth is an amazing miracle in the solar system. It is critical for human beings to understand the rule and importance of life

1

and to protect the surroundings that grow living things on earth.

On my junior high school age, I felt something mysterious and could speculate a law governing the physical substances such as stones, metals, air, and water. The physical substances always show their nature anytime everywhere reproducibly. So I could speculate a common theory behind the phenomena. On the other hand, it is nonsense for me to speculate a simple and reasonable rule of life when I saw a lot of different kinds of species and complicated behaviors of living things at my junior high school days. When I entered college, I decided to study life science especially biophysics and biochemistry, and I studied life science as a professional researcher for more than thirty years. On my age around fifty, now I understand a beautiful and reasonable physical law of life that governs all the living things created from atoms on earth. Because the whole cosmos are created from atoms, it is no doubt that the law of life should govern living things not only on the earth but also in the cosmos.

How the living things are created from atoms? When the secret of creation of life is unveiled, we may feel awe on the mystery of life.

At home in Kasukabe

March 27th, 2015

Hiroyuki Aizawa

Contents

Forewords 1

Contents 3

Chapter 1   Atom and Molecule   4

Chapter 2   Chemical Bond   7

Chapter 3   Water   17

Chapter 4   Carbon Compound   25

Chapter 5   Nitrogen Compound   34

Chapter 6   Catalyst   40

Chapter 7   Peptide   44

Chapter 8   Protein   52

Chapter 9   Nucleic Acid   54

Chapter 10   Gene   58

Chapter 11   Membrane   64

Chapter 12   Cell   71

References 77

Acknowledgements 78

Chapter One
Atom and Molecule

Atom as an ultimate particle for chemical substance

We define an ultimate particle of substances, which could not be divided by any chemical reaction, as an atom. We also call the species of atom chemical elements. At present, scientists discovered and also artificially created 112 elements in the cosmos. All the chemical substances are made from atoms in the periodic table of elements, which are united by chemical bonds. Chemical bonds are created or disrupted by some chemical reaction to change the chemical nature of substances, while the number and element of atoms do not alter at all during the chemical reaction.

We describe the chemical formula of a substance using the number and element of atoms. For example, hydrogen gas exists as a dimer of hydrogen atoms. We describe the chemical molecule of hydrogen gas as H·H or shortly H_2 that is called a chemical formula. Also, we can describe the oxygen molecule and water molecule as O_2 and H_2O, respectively. Furthermore, we can describe the chemical reaction using a chemical formula. For example, we describe the production of water from hydrogen and oxygen gases as $2H_2 + O_2 \rightarrow 2H_2O$. A chemical formula describe that the chemical reaction processes from left molecules to right ones indicating the causal relationship of chemical phenomena. Although the chemical formula of substances is different on the right and left side of the equation, the total number and types of element of atoms do not change during the reaction. Atom is a fundamental constituent of chemical substances and could not be created or disrupted by any chemical reactions. Atom is an ultimate particle of chemical substances.

Molecule as a minimum unit of chemical substance

Chemical substances could be divided into pieces as small as possible to be a minimum unit, a molecule, whose chemical nature is not kept as it is if divided further. You can divide a molecule into ultimate particles at least transiently by chemical reaction, but chemical nature of the original molecule is not preserved in the divided particles anymore. For example, water could be divided into pieces as small as a single molecule H_2O. You can further divide the water molecule by electrolysis into hydrogen and oxygen, while those gases possess a quite different nature from that of water. An atom is an ultimate particle or element, and usually does not exist by itself. Atoms always exist as a molecule or the united atoms of elements. Consequently, the aim of chemistry is studying on the nature of molecules made by unification of atoms.

Philosophers and scientists proposed the concept of a molecule thousands of years ago, but it is quite difficult to prove the existence of a molecule by experiments in the strict sense of word. We do not obtain a definite proof of existence of a molecule until the middle of the 20th century. In 1953, a nucleic acid, a biopolymer with high molecular mass, was proved to be a gene, which is included in a haploid cell as an exactly single copy. Furthermore, if the nucleic acid is divided or chemically modified, the cell changes its nature from that of an original cell, indicating the existence of a molecule of the nucleic acid as a gene in a living cell. Nowadays, we call the biology of nucleic acids or genes as Molecular Biology because of the historic triumph for proof of a molecule from a study of nucleic acids.

Chapter Two
Chemical Bond

Nature of Chemical Bonds

A minimum unit of chemical substances, a molecule, is made from unification of atoms by chemical bonds. A chemical bond is not the existing material but a state of association among atoms, and the bond is created and disrupted during chemical reactions. What is the nature of chemical bonds? Chemical bonds include potential energy. Generally, dissociation of atoms in a molecule requires energy supply by heat, light, or microwave from outside. The dissociated atoms create a new molecule by re-association to form new chemical bonds. Decrease and increase in the internal energy of substances results in an exothermic and endothermic chemical reaction, respectively. In order to start a chemical reaction, an external energy should be supplied into the original substances for disruption of pre-existing chemical bonds whether the reaction is exothermic or endothermic. And upon a re-association of separated atoms into productive substance, the free energy is released to outside from the system. These behaviors of physicochemical energy during a chemical reaction indicate that atoms are associated by force that is attractive to each other in a molecule.

Magnetic force playing a leading role in chemical bonds

Magnets in atoms attract each other in order to form a chemical bond in a molecule. There are several types of chemical bonds classified by the magnitude of magnetic force and stability of chemical association among atoms. Let us study the nature of magnetic force of each element in the periodic table. It is possible to explain that the atomic magnet is an electromagnet, which is derived from movement of an electron around an atomic nucleus. However, here we hypothesize that each atom contains permanent magnets for simplicity. Based on the permanent magnet hypothesis, each chemical bond represents the association of magnets between atoms. The group N in the periodic table should possess n magnets. For example, sodium atom of group I possesses one magnet, and magnesium atom at group II possesses two magnets. Since each permanent magnet has its shape and volume as its nature, the magnets could not share a space within an atom and should repel to each other by a steric effect, which plays a central role in dispersive positioning of each magnet within an atom. Accordingly the two magnets cross at almost right angles to each other by the steric hindrance, while magnetic force tends to form an antiparallel pair of them. As a result, the each magnet could attract a magnet of another atom to form a chemical bond. Aluminum and carbon atoms at group III and IV possess three and four magnets, respectively. A nitrogen atom at group V possesses five magnets, and two of the five magnets align almost anti-parallel and cancel magnetic force to each other inside the nitrogen atom resulting in the three magnets available for chemical bonding. An oxygen atom of group VI possesses six magnets, four of which make two pairs inside the atom, and two of the six magnets are available for chemical bonding. A fluorine atom of group VII possesses seven magnets, six of which make three pairs inside the atom, and one of the seven magnets are available for chemical bonding. A neon atom of group VIII possesses eight magnets, all of which make four pairs inside the atom, and exists quite stable as a single atom molecule with no magnet available for chemical bonding. The periodic table represents the periodic nature of magnetic behavior in elements.

In some textbook of chemistry, an electron or a magnet in the atom is represented as a single dot surrounding a symbol of an element. In such a manner, hydrogen gas or H_2 is described as H:H. Each hydrogen atom has a single magnet, which is represented as a dot in the chemical formula. And the two magnets, each of which belongs to one of the two hydrogen atoms in the gas, make a pair between the hydrogen atoms to create the chemical bond. This is a very simple and straight forward notation to understand the structure of chemical molecule. Now we are ready to go to the next step to understand the species and its nature of chemical bonding based on the magnetic

9

law of the periodic table of elements.

Ionic bond

An ionic bond, which is common in a salt molecule, is relatively unstable in a water solution. Sodium chloride and potassium hydroxide are typical ionic compounds. The ionic bond is created between elements belonging to group I to III and group V to VII. An element in the group I contains a single magnet. On the other side, an element in the group VII contains seven magnets, six of which make three pairs of magnets leaving one magnet available for chemical bonding. Consequently, a sodium atom and a chloride atom can make a molecule Na:Cl with a single chemical bond called an ionic bond.

Crystals of sodium chloride is easily dissolved in water solution immediately. This behavior of sodium chloride depends on the instability of an ionic bond in water solution resulting in dissociation of the two atoms. Because the ionic bond of NaCl is unstable in water, sometimes sodium and chloride atoms react with water molecule to make sodium hydroxide and hydrochloric acid. By this reaction, sodium chloride crystal is dissolved in water quickly. Sodium hydroxide and hydrochloric acid are called base and acid, respectively, and the base and acid are quickly neutralized to produce water. Consequently, the hydrochloric acid and sodium hydroxide, which are transient intermediates of water-solubilizing sodium chloride, neutralize to each other immediately after dissolved as described below.

NaCl (crystal) + H$_2$O → NaOH (aq) + HCl (aq)

NaOH (aq) + HCl (aq) ⇔ NaCl (aq) + H$_2$O

During these reactions, crystal sodium chloride is dissolved quickly into water solution as a solute. Since sodium chloride dissolves in water so quickly, the speed of the first reaction is pretty fast. Once sodium chloride is dissolved in water, sodium chloride keeps equilibrium with hydrochloric acid and sodium hydroxide as described in the second equation.

Although hydrochloric acid is so volatile, we do not detect any vaporized hydrochloric acid from sodium chloride solution and sodium chloride solution does not become alkaline by evaporation, suggesting that the second chemical equilibrium shifts to the right side so much. For example, 1 M NaCl solution contains 1 M NaCl, 55 M H$_2$O, 10^{-7} M HCl, and 10^{-7} M NaOH at pH 7 with a dissociation constant Kd = 1.8 x 10^{-16}.

It is sometimes misunderstood that a sodium atom and a chloride atom are dissociated and exist stably as so-called ion particles in water solution. It is true that chemical bonds should be dissociated temporarily during a chemical reaction, and sodium and chloride should exist as lonely atoms even in a very short period during the reaction. It should also be noted that we identify positively or negatively charged gas-phase ions in mass spectroscopy.

However, almost all the sodium chloride molecules dissolve in water as a solute whose chemical formula is NaCl.

Covalent bond

A covalent bond is an interatomic connection, which is stable in water solution. A typical example of a covalent bond is the stable connection between hydrogen and oxygen atoms in a water molecule. Also, almost all the hydrogen, carbon, nitrogen, and oxygen atoms are connected by covalent bonds in organic chemical compounds. Atoms in groups IV, V, VI in the periodic table and hydrogen atom usually make a covalent bond. In a methanol molecule CH_3OH, a carbon atom binds to three hydrogen atoms by covalent bonds, and also binds to an oxygen atom in a hydroxide group by a covalent bond. The oxygen and hydrogen atoms in the hydroxide group also bind to each other by a covalent bond.

We call those stable bonds covalent ones because the two atoms share a stable pair of valences connected by attractive force between magnets, each of which belongs to one of the two atoms. Generally, disruption of a covalent bond requires much more energy than that of an ionic bond.

Hydrogen bond

Hydrogen is the smallest element whose atomic number is one. Since the atomic magnet of hydrogen is just half of those of other atoms, the hydrogen atomic magnet could neutralize only half of the magnetic field of magnet of atoms other than hydrogen by a magnet pair formation. Consequently, the residual magnetic field could associate with a magnet of other atoms weakly. We call the bond a hydrogen bond since hydrogen plays a key role in creation of this bond. Among the hydrogen compounds, a hydrogen atom in hydroxyl and amine groups is well known to make a hydrogen bond, because magnets in oxygen and nitrogen atoms are not fully neutralized by the covalent bonding with the hydrogen atoms. The residual magnetic field could form a weak hydrogen bond such as O:H···O:H, N:H···O:H, N:H···O=C and so on, where hydrogen bond is represented by a dotted line. On the other hand, saturated alkyl group rarely form hydrogen bond because four hydrogen atoms at all the apices of a regular tetrahedron neutralize all the magnets in carbon atoms completely intra-atomically.

The hydrogen bond is a weak association between atoms intra-molecularly and inter-molecularly. For example, deoxyribonucleic acids form antiparallel double-strands by a lot of hydrogen bonds between bases. Water molecules make hydrogen bonds among them in solution. Hydrogen bonds affect the melting point, boiling point, and viscosity of water as discussed in the next chapter.

Van der Waals bond

Van der Waals bond is a weak attractive association

between molecules such as air molecules in a real gas model. Although the residual magnet force is not so strong as of hydrogen bond, some chemical bonds performs a trace amount of residual magnetic force even after magnet paring mainly by steric effects. The residual magnetic force gives a molecule weak permanent magnetism. The permanent magnetic molecules induce magnetic nature in adjacent paramagnetic molecules and associate with each other. This association is called a Van der Waals bond. A permanent magnet produces a magnetic field, and its strength is rationale to the distance raised to the power of three. The strength of the induced paramagnetic field is also rationale to the distance raised to the power of three. Consequently, the attractive strength of the Van der Waals force is rationale to the distance raised to the power of six. The Van der Waals force keeps a weak association of molecules, but is not strong enough to create a stable bond between atoms to form a molecule.

It should be noted that closed circular electricity in a molecule induces a repellent force to the other electric molecules by electromagnetic induction, while static electricity could attract a dielectric molecule as in the case of permanent magnets. Van der Waals bond suggests that the molecule have a nature of permanent magnetism and/or static electricity but not that of electric current in it.

Hydrophobic bond

Hydrophobic molecules are insoluble to water, and thus tend to gather each other especially in water solution. This type of bond is called hydrophobic bond. This bonding is formed by a weak force that attracts hydrophobic molecules to each other, and passively enhanced by the positive association of hydrophilic molecules such as water molecules by hydrogen bond. A hydrophobic bond forms a weak interaction of molecules, but is not strong enough to create a stable molecule by unifying atoms.

Metallic bond

Metal tends to form a molecule by unification of an indefinite number of metallic atoms. This kind of a bond among metallic atoms, which is typically observed in metals, is called a metallic bond. Stability of metallic bonds in water solution depends on the type of elements, and solubility of metal to water is called ionization tendency. Typical metallic atoms contain one to three dipole magnets, each of which make a pair of two magnets between atoms as a metallic bond. The magnetic pair in a metallic bond is not so stable as in a covalent bond, and the pair is often dissociated. In metals, however, one metallic atom is surrounded a lot of metallic atoms, each one of the two dissociated magnets finds a new partner to make a magnetic pair easily and quickly. Since metallic magnets are able to rotate freely around the atom to make a pair with a magnet of neighbor atoms, we call it a free magnet. A free magnet rotates freely around the atom to make a metallic bond with neighboring atoms freely. Consequently, it is a nature of metallic bonds that any magnet of metallic atoms does not have its steady partner, and all the atoms appear to share their magnets in metal just like a benzene ring carbons that share six antiparallel magnets as shown in chapter four.

Chapter Three
Water

Water as a source of life

Water is a fundamental material for life. It is a solvent for life substances, and also a source of hydrogen and oxygen, that is, essential elements of life molecules. It is no exaggeration to say that life is born from water. Understanding the nature of water leads us to the secrets of life. From this point of view, I selected water as the first molecule of life in this book. How do atoms interact with each other by chemical bonds in a water molecule?

Water is a chemical compound created from just two elements, that is, hydrogen and oxygen. Water is a solution under 1 atm at room temperature, and frozen to become a solid ice under 0 degrees. And water becomes a steam gas above 100 degrees when boiled. Solid, liquid, and gas are called the three states of matter. Solid water or ice is known to have several forms of crystals. Within those crystals, water molecules are aligned regularly by intermolecular hydrogen bonds. Scientists have revealed some secrets of a hydrogen bond by investigating the structure and function of ice crystals. When boiled, water vaporizes to colorless and transparent steam gas by absorbing the latent heat of vaporization. Although some argument exists among scientists on molecular structure of gas, we should go ahead without serious discussion on this issue.　Here we go to the next step to study on the nature of water solution, which is an essential material for life.

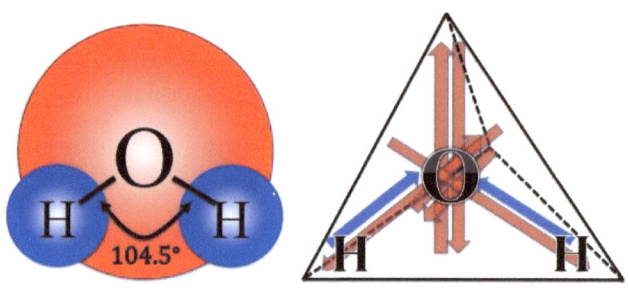

A model of water molecule (Figure 1)

How does the water molecule exist in water?　At first, let us see the structure of a single water molecule.　Water has an oxygen atom at the center of the molecule, and two hydrogen atoms bind to the oxygen atom by covalent bonds symmetrically.　The oxygen atom has six magnets, four of which form two antiparallel pairs of magnets and are consequently neutralized magnetically.　Each of the residual two free magnets in the oxygen atom makes a covalent bond with a magnet of a hydrogen atom.　The two covalent bonds form an angle of 104.5 degrees, and hydrogen atoms are set at two of the four apices of a regular tetrahedron, at the center of which the oxygen atom locates.　Two pairs of magnets locate at the residual two apices of the regular tetrahedron.　Taken all together, water molecule forms a tetrahedron structure with an oxygen atom setting at the center. Because the magnetic dipole moment of hydrogen is half of that of a standard atomic magnetic dipole moment, the covalent bond between hydrogen and oxygen atoms retains half of a standard atomic magnetic dipole moment. Among the half dipole moment, one sixth of standard dipole moment locates on the hydrogen side, and two sixths or one third of standard dipole moment locates on the oxygen side. According to an effect of the residual dipole, water molecule could make a weak bond with other molecule. This is the explanation of the mechanism of hydrogen bond by the magnetic hypothesis of chemical bonds. The magnetic hypothesis, which has some merits to include a steric effect on the chemical bonds locating in the three-dimensional space, is expected to contribute to development of biophysics and biochemistry from now on.

Next, let us see an effect of a hydrogen bond on an interaction between water molecules. Flowing water has no fixed shape, and thus we could scoop up water with a cup. In other words, we could divide water between any molecules. Also, water molecules change its position to each other almost freely. On the other hand, water molecules attract with each other, and the strength of the attractive force could be measured as a tension of water surface. This attractive force is produced by a hydrogen bond between two water molecules. Hydrogen bonds play a central role in viscosity of water.   Below the 0 degrees, water loses its fluidity and is frozen as an ice to form a solid matter, in which a water molecule is tightly connected to two water molecules by two hydrogen bonds.   This is the frozen point of the water.   Ice is the water whose molecules are all connected to each other by hydrogen bonds.   The rigidity of ice tells us how the hydrogen bond is rigid and strong once it is formed.   In order to disrupt a hydrogen bond of water molecules, we need to warm the ice above 0 degrees.   This is the melting point of the ice, and it tells us how hydrogen bond plays a role in the three states of water.

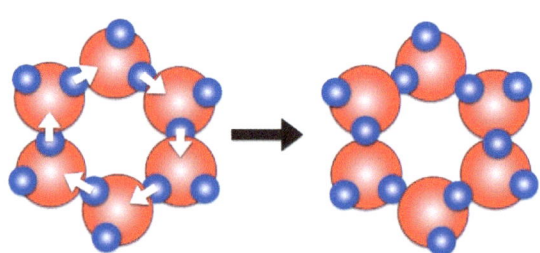

A hydrogen transfer chain reaction in water (Figure 2)

Next, let us see a natural chemical reaction among water molecules in liquid water.   The hydrogen atom is not permanently fixed in a single molecule in water solution. Rather, the hydrogen atom is occasionally exchanged with that in other water molecules at a frequency of fourteen times per second. When released from an oxygen atom of a water molecule, a hydrogen atom tries to join its hand with next oxygen. Almost all the released hydrogen atoms, however, return to join its old partner of oxygen. The released hydrogen atom rarely binds to next oxygen atom that has released a hydrogen atom prepared for joining its hand with next hydrogen. It should be noted that the hydrogen bond is not corresponding to the covalent bond between hydrogen and oxygen during the hydrogen transfer reaction. Consequently, an oxygen molecule should rotate to join its hand with next hydrogen during the transfer reaction. The released next hydrogen atom should bind to next-next oxygen and so on. This chain reaction does not finish until the very first oxygen atom, which released the very first hydrogen atom to start this chain reaction, joins its hand with the very last hydrogen released from the final oxygen that has joined with next hydrogen atom.

Usually, the hydrogen transfer chain reaction occurs only when water molecules take a position to form a closed circle, whose geometric morphology is suitable for the chain reaction, as shown in the figure 2 representatively. Consequently, all the chain reactions are performed almost at once. Even a single misplacement of an atomic angle in the circle of water molecule makes the chain reaction impossible. A bond between hydrogen and oxygen between water molecules vibrates at a quite high frequency at around 10^{12} Hz while the frequency of a hydrogen atom transfer among water molecules is around $10^1 - 10^2$ Hz, suggesting that the hydrogen transfer chain reaction occurs quite rarely in water.

When water is electrolyzed, the hydrogen transfer chain reaction occurs not in a closed circular manner but in a linear manner between a cathode and an anode. As everybody knows, however, pure water does not conduct electric currents as an insulator because a water molecule bends in its center like a wedge resulting in a closed chain reaction rather than a linear one as shown in Figure 2. The hydrogen transfer chain reaction tends not to be performed on a linear path. However, only a slight amount of salt in water solution dramatically increase a conductance of the solution as an electric catalyst because of a doping effect of salt, whose ionic bond is quite frequently dissociated at any angle catalyzing hydrogen transfer to an anionic atom and production of next hydrogen by cationic trap of hydroxyl group from a water molecule. Consequently, low concentration of salt dramatically decreases electric resistance and makes water solution much conductive.

At a glance, water in a glass does not appear to perform a chemical reaction by itself. However, hydrogen transfer chain reaction occurs fourteen times per second on every water molecule in a cup. This mysterious nature results in the diamagnetism of water. The closed electric flow induces magnetism, and the change of magnetic field induces electric current in the next chain reaction. In such a way, the hydrogen transfer chain reaction caused an electromagnetic wave along the central axis of the chain reaction circle.

When applied to water solution, the electromagnetic wave induces electric voltage corresponding to amplitude of the wave, and a current flows upon the electric discharge. The electric voltage and current correspond to the dissociation of hydrogen from oxygen and transfer of the hydrogen to next oxygen, respectively. Just after transfer of hydrogen as electric currents, the hydrogen atom oscillates within the next water molecule for a while to increase its temperature. In other words, a high frequency electromagnetic wave could increase the temperature of water solution. This is exactly the mechanism by which a microwave oven heats up water solution in a kitchen. Although an ice is made of water molecules, the ice molecules are rigidly fixed to each other by hydrogen bonds and could not rotate for hydrogen transfer, resulting in a high resistance of ice. Consequently, the ice is an insulator and microwave oven does not warm ice efficiently. On the other hand, low concentration of salt increases conductance of the water dramatically and a microwave oven heats up a salt solution more efficiently than pure water.

A water molecule H_2O is quite simple as a chemical substance. However, just an aggregate of such a simple molecule, water, demonstrates various complex physicochemical and electromagnetic phenomena of materials. The physicochemical nature of water plays a fundamental role in life activities, and various life molecules are made from water as a source of hydrogen and oxygen such as carbohydrates, peptides, and nucleic acids as described in the following two chapters.

Chapter Four
Carbon Compound

Chemical bonds of a Carbon atom

The sixth element of the periodic table, carbon, belongs to the group IV harboring four magnetic bars in an atom. Each of the four magnets locates at a top of a regular tetrahedron as a dangling bond. The magnetism of the carbon atom is exactly zero because of the perfect space symmetry of the regular tetrahedron of the carbon atom. This symmetric distribution of four magnet bars is so stable and almost fixed tending not to change its location and angle to each other in a carbon atom. As a result of this stability of chemical bonds, a carbon atom becomes a center of chirality when a carbon atom binds to four compounds different to each other. The four chemical bonds in a carbon atom are rigid and fixed to each other. Consequently, high molecular mass organic compounds take a rigid and characteristic three-dimensional structure.

Organic compounds as life materials

Except for carbonic dioxide and carbonic acid, all the compounds containing carbon atoms are called organic compounds. Among the organic compounds, a molecule made of carbon, hydrogen, and oxygen by covalent bonds is called a carbohydrate. Green plants produce carbohydrates by photosynthesis from water and carbon dioxide in chlorophylls. Carbohydrates are fundamental life materials, from which various kinds of life materials such as nucleic acids and lipids are created.

Carbon is a fundamental element for life materials. Carbon gives the nature of stability to high molecular life materials. In order to make a high molecular compound by covalent bonds, each atom should have at least two chemical bonds. The elements belonging to the group II/VI and III/V has two or three chemical bonds in a single atom, respectively. Although metallic atoms of group II/III can make a large solid molecule by metallic bonds with free magnets, pure crystal metals are not typical in living things. Actually, almost all the metals exist as minerals and protein complex in living things. On the other hand, elements belonging to the group V/VI with some intra-atomic magnetic pairs are relatively reactive and thus could not make a large molecule by themselves because of the chemical instability. For example, the following chemical compounds are not stable in general.

$NH_2 \cdot N=N \cdot N=N \cdot NH \cdot NH \cdot N=NH$

$HO \cdot O \cdot O \cdot O \cdot O \cdot O \cdot O \cdot O \cdot O \cdot O \cdot O \cdot OH$

In water, those compounds immediately react with water molecules resulting in production of low molecular mass compounds.

On the other hand, a carbon atom locates its four magnetic bars at the vertexes of a regular tetrahedron symmetrically. Two of the four magnets in each carbon atom could bind to next carbon atoms to form a high molecular polymer stably.   For example, a heptane molecule consists of seven carbon atoms, which are polymerized by carbon-carbon covalent bonds as below.

CH_3-CH_2-CH_2-CH_2-CH_2-CH_2-CH_3

Heptane is stable on a surface of water solution.   Heptane is a hydrocarbon included in gasoline performing a hydrophobic nature as an organic solvent.   Heptane is not solved in water at all, and floats on the surface of water as oil.   If one hydrogen atom binding to every carbon is substituted with hydroxyl group, we call the compound a carbohydrate.   Because of the hydrophilic nature of hydroxyl group, a carbohydrate is quite soluble and stable in water solution.   Even in the presence of water, life could not be created without carbon.

Sugar and its polymer

Sugar is a typical compound of carbohydrates.   Plants perform photosynthesis in its chloroplasts to synthesize glucose from water and carbon dioxide by carbonic fixation.   Therefore, glucose $C_6H_{12}O_6$ is the first organic compound made from natural inorganic compounds, water and carbon dioxide.   The chemical formula of glucose is described as below.

CHO-$CHOH$-$CHOH$-$CHOH$-$CHOH$-CH_2OH

The molecular mass of glucose is 180, which is ten times larger than that of water molecule.   This monosaccharide is a unit material of high molecular mass substances such as starch and glycogen, which are biomaterials for energy storage. High blood glucose induces diabetes, one of three lifestyle-related illnesses. A simple organic compound, sugar, is an important carbohydrate that plays a central role in human health.

Glucoses are dehydrated for condensation to produce polymers such as cellulose, glycogen, and starch. Cellulose is a major component of plants. Cellulose is a linear polymer of glucose, and several polymers align in parallel with cross-bridges of hydrogen bonds between polymers. Because of the cross-bridge structure, cellulose is hard enough to support a big tree taller than 10 meter high. Cellulose is a source of paper, and also delicious foods for microbes and animals, which can digest cellulose. Carbohydrates are fundamental organic compounds for all of the living things.

Double bond between carbon atoms

Carbon atoms bind to each other by three types of chemical bonds, a single (C-C), double (C=C), and triple (C≡C) bonds according to the number of magnet pairs between carbons within a chemical binding. The carbons involving in the double and triple bonds are called unsaturated carbons. Since the triple bond is made by an expansion of the double bond formation, let us see the physicochemical nature of the double bond in detail here.

All the chemical bonds between carbon atoms are formed primarily by a single bond, which make a pair of magnets between carbon atoms just along the axis of bonding. One among a double or triple bonds is this type of a single covalent bond. The second bond could not be formed on the axis between carbon atoms because of the steric hindrance effects of magnets, and consequently the second magnet of both atoms form a certain angle to the axis of the binding between carbons. Here we hypothesize that the angle is 90 degrees, which is typical for an ideal double bond. Since opposite poles of magnets attract with each other, two of the second magnets bind laterally between carbon atoms at an inter-atomic interval forming an anti-parallel pair. This is a secret of the double bond. A triple bond adds one more anti-parallel pair of magnets, which form an angle of 90 degrees with a single and double bonds, respectively. The triple bond magnets also bind laterally between carbon atoms at an inter-atomic interval forming an anti-parallel pair.

Based on this magnetic interaction mechanism of chemical bonding, it is quite natural that the second unsaturated bonds has a distinguished character from that of the first covalent bond formed along the axis of the two carbon atoms. The two magnets forming the second bond are set apart at an inter-carbonic interval and perform an antiparallel interaction laterally in a plane which includes the first bond axis, resulting in a weak interaction. The weakness of the second bond suggests that the second bond is easy to be broken and ready for various chemical reactions.

Among four magnets in a carbon atom that forms a double bond, the first magnet locates along the bond axis and the second one stays across the axis forming an angle of 90 degrees. The other two magnets in the atom locate to minimize the effect of steric hindrance. In other words, the two magnets locate in a plane, which includes the first bond axis, and form an angle of 120 and 90 degrees with the first and second magnet, respectively. After all, the four covalent bonds extending from the two unsaturated double bonding carbons locate symmetrically in a plane to form an angle of 120 degrees with the bond axis of the two carbons.

Conjugated double bond and benzene ring

If double bonding carbons were connected in series by single covalent bonds, we call it a conjugated double bond. One example of conjugated double bond compounds is described below.

$CH_2=CH \cdot CH=CH \cdot CH=CH_2$

When the very left and right side carbons are connected covalently, it is called benzene ring.

The conjugated double bond is not only the simple series of single and double bonds but also shows an amazing nature that all the atoms of the molecule locate on a single plane. As mentioned above, four single covalent bonds extend from double bond carbons in a single plane. In conjugated double bonds, the next double bond also makes another plane by itself. The plane, however, could not rotate freely around the single bond that connects two double bond carbons, because each of the second magnets in conjugated double bond carbons stand perpendicular to the plane and forms antiparallel with the magnet of the next carbon atom. As a result, all the carbons are unified by attractive force among the antiparallel magnets.

For example, benzene ring is described as below by chemical formula.

·CH=CH·CH=CH·CH=CH·

The very left and right carbons are connected to form a hexagon. If the second magnet in the double bond is indicated by arrow, the benzene ring is described as below.

· ↑ = ↓ · ↑ = ↓ · ↑ = ↓ ·

Here, we do not ask the north and south identity of a pole of arrow. Since the left and right carbons are united by a single covalent bond to make a hexagon, actually each of the second magnets in the benzene ring bind to next one on both sides equally. Consequently there is no difference between single and double bonds in the chemical formula of the benzene ring.

Double bond between carbon and oxygen

Double bond is formed not only between carbon atoms

but also between carbon and oxygen. While the magnetic mechanism to form a double bond is common among all the double bond, its chemical nature depends on the elements in the bond. Here, we study on nature of a double bond formed between carbon and oxygen, that is, carbonyl group, whose chemical formula is R-(C=O)-R'.

In the carbonyl group, carbon and oxygen atoms form a covalent bond and also a lateral association of the second magnets of the both atoms, which form an angle of 90 degrees with the covalent bond axis as in the case of unsaturated double bonds between carbon atoms described above. Consequently, remaining two magnets in the carbon atom form respectively an angle of 120 degrees from the covalent bond in a plane perpendicular to the second magnet, and remaining four magnets in the oxygen atom make two pairs, which form respectively an angle of 120 degrees to the double bond axis. The laterally associated magnets have residual weak magnetic force around the bond and thus highly reactive. While a double bond carbon is associated with three atoms around it, double bond oxygen is associated with a carbon atom and all the other space is open for access of reactive agents in a solution. In other words, an oxygen atom that binds to carbon atom by a double bond is more reactive than double bonding carbons. For example, an amine group could react with the double bonding oxygen as described below.

R-, R'-C=O + H_2-N-R" →R-, R'=C-OH + HN=R"

Carbonyl group is an important reaction group of organic compounds together with unsaturated carbons. Carbonyl and amine groups play a pivotal role in protein folding and double strand DNA structure, which are two essential and central materials for life activities.

Chapter Five
Nitrogen Compound

Valance of Nitrogen

Nitrogen is the seventh element in the periodic table, belonging to group V. A Nitrogen atom has five magnets, two of which make a stable pair in the atom, while the other three magnets locate symmetrically to minimize the effect of steric hindrance. Accordingly, valance of nitrogen is expected to be three or tetravalent. Actually, ammonium consists of one nitrogen atom and three hydrogen atoms. The molecular formation from the four atoms is explained almost the same way as that of water molecule based on the magnet hypothesis. In an ammonium molecule, three hydrogen atoms locate at the apex of a triangle, at the center of which a nitrogen atom locates. A pair of magnets locates at the center of the triangle and become perpendicular to the plane of the triangle.

Some scientists expect that the three hydrogen atoms locate at the three apexes of a regular tetrahedron, at the center of which a nitrogen atom locates. To tell the truth, it is known that the magnets could move to each other within a nitrogen atom to some extent, and thus it might be better to say that the positions of hydrogen atoms are flexible and not fixed rigidly in the nitrogen atom. This is consistent with the fact that a nitrogen atom does not create a chiral molecule as a carbon atom does. Because of the flexibility of nitrogen valance, nitrogen compounds could form flexible high molecular materials and thus play a critical role in enzymatic activity of proteins and double strand formation of nucleic acids.

It is obvious that nitrogen is trivalent in an ammonium molecule, while some nitrogen atom is pentavalent. For example in nitric acid HNO_3, a nitrogen atom binds one hydroxyl group with a single bond and two oxygen atoms with double bonds. In nitrous acid HNO_2, however, a nitrogen atom binds one hydroxyl group with a single bond and an oxygen atom with a double bond. Since oxidation of nitrous acid produces nitric acid, a single nitrogen atom could be trivalent or pentavalent during an oxidative chemical reaction. This fact indicates that all the five magnets of a nitrogen atom could contribute to chemical bonding depending on conditions. In other words, a pair of magnets in a nitrogen atom could dissociate and contribute to a chemical bond. The intra-atomic magnet rearrangement required the strongest force in chemical reactions.

We learned trivalent and pentavalent nitrogen, and sometimes nitrogen also appears to have four chemical bonds in a molecule such as so called an ammonium ion. This ion, however, forms an ionic molecule and for example an ammonium and chloride ions produce ammonium chloride NH_4Cl. In the molecule, a nitrogen atom has five bonds, one of which is an ionic bond that is instable in water.

Amine and amino acid

Amine is a derivative of ammonium, whose hydrogen atoms are substituted by alkyl groups. Chemical formula of a primary, secondary, and tertiary amines are $R\text{-}NH_2$, $R\text{-},R'\text{-}NH$, $R\text{-},R'\text{-},R''\text{-}N$, respectively. All the amine compounds include a trivalent nitrogen atom. Amine nitrogen has a loose pair of magnets and could make an ion molecule by dissociation of the pair to make an ion bond with highly reactive molecules. On the other hand, the amine trivalent nitrogen could make a double bond with alkyl group by releasing hydrogen atom. The high reactivity is the nature of amine compounds.

Amino acid is a primary amine derivative, whose α-carbon binds to carboxyl group or $\text{-}COOH$. Amino acids are organic acidic compounds with molecular mass of around a hundred, and major constituents of proteins. Two amino acids could form a peptide bond $R\text{-}CO\text{-}NH\text{-}R'$ by dehydrating polymerization between carboxyl and amine groups. Two or more amino acids produce a peptide by a linear polymerization by peptide bonds. Especially, a high molecular mass peptide is called a polypeptide, and polypeptides are assembled by hydrophobic interaction or cross-linked by disulfide bonds to form a protein. Some protein is a highly specific catalyst, called an enzyme.

Protein is an important and central molecule for a cell to perform life activities.

Base of nucleic acid

Purine and pyrimidine are ring-shaped organic compounds that contain nitrogen atoms. Purine and pyrimidine form a single plane ring structure by conjugated double bonds consisting of unsaturated carbons, carbonyl groups, and amine groups. In order to understand the conjugated double bonds of unsaturated carbons and carbonyl groups, please go back to the previous chapter. Two lonely and a pair of magnets of tertiary amine nitrogen, which makes a double bond with carbon in a base, forms a triangle in a single plane. Residual lonely magnet stands up perpendicularly from the plane. This magnet forms conjugated double bonds with unsaturated carbons and carbonyl group carbons. Also, secondary amine nitrogen in a ring of a base stays in the same plane as conjugated carbons by support of the base ring structure even when the nitrogen is not conjugated. Consequently, the base ring takes a flat structure within a single plane, and thus makes two or three hydrogen bonds between another base of the complementary strand in nucleic acids.

An amino acid also has a carboxyl and primary amine groups, and the two groups sit side by side in a peptide bond. As will be discussed in detail later, four atoms of carbonyl group and secondary amine locate in a single plane.   This plane plays an important role in protein folding to form secondary structure such as α-helix and β-sheet.   Only four elements, that is, hydrogen, carbon, nitrogen and oxygen create a carbonyl amine, which plays a central role in protein and DNA, two essential materials for life activities.   This simple and sophisticated beautiful secret of living creature is nothing other than the master peace of God.   It is also true that the four elements are essential to make basic pieces of life materials not only on earth but also in the cosmos.

Plant alkaloids

We know amazingly many nitrogen compounds other than amines, amino acids, and bases.   Most of nitrogen compounds are identified in plant, and historically they are called plant alkaloids.   The common nature of alkaloids is a compounds including nitrogen.   Various alkaloids have various characters of chemical reactions, and perform their own specific physiological activities.   Alkaloids could be classified into subcategories based on the nature of their molecular structures, while we still do not have a clear classification system.   Among the alkaloids, we call a nitrogen-containing heterocyclic compound a true alkaloid.

Various physiological activities of alkaloids play a role in human history as medicine and poison from an ancient period.   This nitrogen compound group plays a very important role in life, and please read some technical books for further study in detail.

Chapter Six
Catalyst

Catalytic action

Generally, a chemical reaction goes faster in higher temperature, which accelerates a molecular motion to make chemical bonds instable. In this case, all the chemical reaction rates increase. There is no specificity for high temperature to accelerate chemical reactions. On the other hand, a ring-shaped chain reaction of hydrogen transfer among water molecules is accelerated specifically for the water molecules included in the ring. An addition of a small amount of salt in water dramatically stimulates the hydrogen transfer reaction in water solution. In this reaction, salt acts as a catalyst, which stimulates the chemical reaction specifically.

Let us show another example for a catalytic reaction. Please image an acetic acid, which is solved in a water solvent. This solution is called vinegar. The chemical formula of an acetic acid solution is described below.

$$CH_3\text{-}(C=O)\text{-}OH + H_2O \rightarrow CH_3\text{-}(C\text{-}OH)=O + H_2O$$

In a water solution, an acetic acid performs an intra-molecular hydrogen transfer reaction from hydroxyl group to carbonyl group. When watching the chemical formula, a water molecule does not appear to play any role in the hydrogen transfer reaction because its chemical composition does not alter at all before and after the reaction. However, if each atom could be distinguished, a hydrogen atom of the hydroxyl group is transferred to a water oxygen, and a hydrogen atom of water molecule is transferred to an oxygen atom of the carboxyl group during the hydrogen transfer reaction. In other words, a water molecule intermediates hydrogen transfer between hydroxyl and carbonyl groups within an acetic acid. Without the water molecule, it takes a large energy for an acetic acid to transfer hydrogen within a molecule. On the other hand, it takes low energy through the intermediation of water molecule just like a closed hydrogen transfer chain reaction in water solution at room temperature. As a result of this transfer reaction, two oxygen atoms in an acetic acid appear to share a single hydrogen atom. In this reaction, a water molecule catalyzes the hydrogen transfer reaction.

As is well known in the catalytic reaction of platinum, which stimulates burning of hydrogen in air, a metal catalyst does not change its atomic structure on its action. On the other hand, an organic catalyst usually intermediate some atoms between substrates and their own atoms as mediators. Since this type of catalyst accepts and donates exactly the same number and kind of atoms from and to substrates, respectively, the catalyst does not alter its chemical structure before and after the reaction.

Catalyst-producing catalyst and self-reproducing catalyst

Some catalyst stimulates a reaction, which produces another catalyst. For example, a protease is an enzyme or a protein catalyst that stimulates hydrolysis of peptide bonds in polypeptides. Some protease is synthesized as a latent inactive form, and is activated after hydrolysis by another protease. Then, the activated protease stimulates a hydrolysis of its substrates. Accordingly, proteases could form a protease cascade to amplify the activation signal of catalyst.

Furthermore, some catalyst is self-reproducing. In other words, some catalyst stimulates a chemical reaction, which produce the catalyst itself. This is actually one of the catalyst-producing catalysts just mentioned above. A typical example of the self-reproducing catalyst is prion. Prion is a protein causing mad cow disease, which stimulates production of prion. Prion is a chaperon, which supports and modifies protein folding to build up a three-dimensional structure. Prion stimulates a reaction that changes conformation of inactive latent prion to an active form. When a catalyst stimulates a reaction producing itself, the catalyst obtains an ability to reproduce naturally. Although prion protein is not a living thing by itself, scientists pay attention to prion as a reproductive polypeptide.

Another typical self-reproductive catalyst is a nucleic acid. Nucleic acids are high molecular mass molecules assembled as antiparallel double strands of deoxyribonucleotides or DNA. While its precise structure and function are discussed later, here we note that replication of DNA is catalyzed by anti-strand DNA, whose replication is catalyzed by the anti-anti-strand DNA or DNA itself. Complementary DNA is synthesized by polymerization of nucleotide using its anti-strand DNA as a matrix or catalyst. Polymerization reaction processes quickly as a condensation reaction using a high energy of phosphate bond in nucleotides, and DNA polymerase enzyme catalyzes the condensation reaction. Without anti-strand DNA matrix, however, polymerization reaction of DNA does not occur even in the presence of nucleotides and DNA polymerase. Anti-strand DNA recruits one of the four deoxyribonucleotides as a matrix to accelerate the polymerization of DNA in an anti-strand DNA sequence specific manner together with DNA polymerase enzyme. A single strand DNA could produce thousands of complementary DNA without any chemical alteration of the DNA molecule, and the synthesized complementary DNA could also produce thousands of its complementary DNA or the original DNA. In that way DNA is a self-reproductive catalyst that could multiply itself almost infinitely. After the success of molecular biology, now we know that the double strand DNA is a genetic material, that is, a secret of life.

Chapter Seven
Peptide

Peptide as an amino acid polymer

As we study in the chapter five, an amino acid is a general term for a primary amine, whose α-carbon binds covalently to a carboxyl group -COOH. Amino acids perform its various properties according to the alkyl groups, which bind to the α-carbon. Theoretically, the number of species of amino acids is expected to be infinity. On the other hand, only twenty kinds of amino acids have been identified as units for polymerization to produce polypeptides in all the living things on the earth. Amino acids are condensed by dehydration between carboxyl and amine groups between two amino acids, forming a peptide bond. This reaction is described by a chemical formula as shown below.

NH_2-CH(R^1)-COOH + NH_2-CH(R^2)-COOH $\rightarrow$ NH_2-CH(R^1)-CO-NH-CH(R^2)-COOH

This dipeptide, which consists of two amino acids, is simply notated by a single or triple letters indicating the type of amino acid. For example, a dipeptide consisting of an aspartic acid Asp and glycine Gly are described as Asp-Gly for three letters notation and DG for a single letter notation. In the same manner as above, peptides consisting of a large number of amino acids are also written in a sequence of single letter or three letters. The

44

sequence starts from an amino-terminal amino acid and terminates at a carboxyl-terminal amino acid in the same way as peptide synthesis in a living cell.

In living things, we find various lengths of peptides from two to thousands of amino acids. All the amino acids are connected by peptide bonds in the peptides. The peptide bond plays a central role in performing the general characteristics of peptides. On the other hand, each amino acid harbors its specific alkyl group called side chains, which play a specific role in various physiological activities. Peptide hormones have various activities, which are necessary for regulation of physiological functions such as appetite, sexual desire, blood pressure, and blood sugar level. Some technical books may describe precise activities of each physiological peptide in detail. On the other hand, we focus on the fundamental characteristics of peptides in order to approach a secret of life.

Structure and function of a peptide

Common and basic structure and function of peptides depend on the nature of peptide bonds. The basic structure of a peptide bond is R-CO-NH-R', where R and R' are some carbon compounds. Carbonyl group takes a plane structure with three bonds, which make an angle of 120 degrees to one another. The secondary amine also takes a plane structure with three bonds making an angle of 120 degrees to one another. The two planes of carbonyl group and amine are connected by a single bond between carbon and nitrogen atoms, and the two planes could rotate freely around the bond.   Consequently, the two planes of carbonyl group and amine are independent and twisted at a condition without regulation.

Generally, peptides are solved in water solution well by two hydrophilic residues, that is, carbonyl group and amine. And the solubilized peptide forms hydrogen bonds with water molecules performing hydrogen transfer chain reaction with water molecules.   Furthermore, a relatively long peptide or polypeptide could form hydrogen bonds within the peptide molecule to form the secondary and tertiary structures.   Here, let us see hydrogen bonds and a hydrogen transfer chain reaction of peptides.   It is a good idea to remember what we learned in case of water molecules in chapter three.

Hydrogen bond between a peptide and water

In a peptide bond, R-CO-NH-R', there are two atoms that could form a hydrogen bond.   They are oxygen and nitrogen atoms.   The oxygen of carbonyl group forms a double bond with a carbon atom, which is represented as R-(C=O)-NH-R' in a chemical formula.   Water molecule hydrogen binds to the carbonyl oxygen atom in the peptide bond.   A hydrogen atom of the secondary amine is also able to make a hydrogen bond with an oxygen molecule in a water molecule.

The next question is whether the hydrogen atom of a peptide bond could transfer to the oxygen atom or not. The answer is yes. Just like a hydrogen transfer chain reaction in water, a peptide hydrogen atom transfers by a closed chain reaction as follows. A hydrogen atom of a water molecule transfers to an oxygen of a carbonyl group in a peptide bond. Consequently, the double bond of the carbonyl group is disrupted to a single bond. Next, the dissociated magnet of a carbon atom forms a double bond with that of a nitrogen atom, and consequently the nitrogen atom releases hydrogen atom to an oxygen atom of another water molecule. In this way, a peptide bond could be a part of a hydrogen transfer chain reaction in water solution.

The chemical equation of the reaction is described as below.

$$H + R\text{-}(C=O)\text{-}NH\text{-}R' + O\text{-}H \rightarrow R\text{-}(C\text{-}OH)=N\text{-}R' + H_2O$$

However, the product, an imine, is unstable and tends to go back to an amide quickly as described below.

$$O\text{-}H + R\text{-}(C\text{-}OH)=N\text{-}R' + H \rightarrow H_2O + R\text{-}(C=O)\text{-}NH\text{-}R'$$

This reversible reaction is expected to occur at a frequency of 14 Hz. The chemical equilibrium largely shifts to the amide formation. A dramatic and important effect of this equation is that all the atoms in a peptide bond locate in a single plane during the reaction. The carbonyl group and amine form each specific plane in an amide, and an imine also forms a plane. Since all planes should be piled up into one in an equilibrium state, all atoms of a peptide bond locate in a single plane by the hydrogen transfer reaction. On the contrary, when all the three planes are piled up into one, the peptide bond is ready for a hydrogen transfer chain reaction in water solution. In other words, the angle and rotation between the carbonyl group plane and amine plane determine and regulate the rate of hydrogen transfer reaction, respectively.

Hydrogen bond within a polypeptide

A polypeptide, which consists of more than ten amino acid residues, sometimes forms hydrogen bonds within the molecule. The intra-peptide hydrogen bonds play a central role in the three dimensional structure of a peptide molecule. How could an intra-peptide hydrogen bond be formed in a polypeptide? An oxygen atom of a carbonyl group and a hydrogen atom of an amine are able to make a hydrogen bond as hydrogen acceptor and donor, respectively. A hydrogen bond could be formed between the oxygen and the hydrogen, each of which belongs to different amide within a single polypeptide. It should be noted that a single hydrogen bond is so weak that the bond is disrupted at a temperature above 0 degrees. If three-dimensional conformation of the peptide is stabilized by ionic or hydrophobic interactions through side residues, even a single or a few hydrogen bonds could be formed stably among amides in a polypeptide main chain. In order to form a stable three-dimensional structure by hydrogen bonds in solution at room temperature, more than ten hydrogen bonds should be aligned in parallel within a polypeptide.

Using hydrogen bonds between amides, a polypeptide forms three characteristic secondary structures, α-helix, parallel β-sheet, and antiparallel β-sheet. In the α-helix structure, polypeptide takes a structure of right- or left-handed helix with hydrogen bonds along an axis of the helix. A hydrogen bond in α-helix participates in a transfer reaction of the hydrogen atom between amine and carboxyl group just like the case between an amide and water molecules as mentioned above. In a single α-helix, three independent strands of hydrogen chain reactions take place along the axis of α-helix. However, the chain reaction is not closed within a α-helix. In order to turn on an electric current as a hydrogen transfer chain reaction in α-helix, it is necessary to add and remove a hydrogen atom on the very first and last carbonyl group and amine of each strand of chain reactions, respectively.

In the β-sheet structure, hydrogen bonds are formed among several polypeptide chains, which align in parallel or anti-parallel manners. Hydrogen bonds bind several polypeptide chains into a sheet-like structure just like weaving. The β-sheet sometimes makes a tubular structure by connecting both ends of the sheet by hydrogen bonds. The number of peptide strands included in the tubule should be more than two in a β-sheet tubule. In the tube of a peptide sheet, hydrogen bonds could perform the hydrogen transfer closed chain reaction within a polypeptide. Namely, the closed chain reaction always takes a place in a ring of peptide bonds, which are connected by hydrogen bonds in a plane perpendicular to an axis of the tube. Moreover, once a closed ring chain reaction occurs, next ring reaction is induced in the opposite direction by electromagnetic induction. Since the equilibrium between amide and imine in a peptide bond largely shifts to amide, once amide is changed to imine by hydrogen transfer, the reverse reaction from imine to amide occurs immediately as an automatic oscillation. This electric oscillation in a closed circular ring transmits in a β-sheet tubule as a transverse wave along its long axis. Because the length of the β-sheet tubule is limited, a standing wave is generated in the tubule by transmission of the electric oscillation. This standing wave could be generated either in a parallel or anti-parallel β-sheet tubule. Once a closed circular chain reaction occurs, the carbonyl group and amine locate in a single plane by an imine double bond, and consequently hydrogen bonds in the β-sheet tubule are stabilized. As a result, the β-sheet tubule is the most stable structure among several peptide conformations. Hydrogen bonds among amides in a polypeptide stabilize its three-dimensional structure by mechanically, electromagnetically, and a resonance effect of the both.

There are two types of β-sheet, that is, parallel and anti-parallel sheets. When we observe a three-dimensional structure of hydrogen bonds between the peptide bonds, N-H bond axis and C=O bond axis stand at a skew position in the parallel β-sheet, while both axes cross at a point in the anti-parallel β-sheet. Generally, hydrogen bond is formed stably at the crossing position, especially on a single line. Accordingly, anti-parallel β-sheet is more stable than parallel one. The β-sheets play a central role in protein folding and maintenance of polypeptide conformation.

The three types of three-dimensional structure of a polypeptide, that is, α-helix, parallel β-sheet, and anti-parallel β-sheet, play a general role in regulation of polypeptide folding and its heat stability.

Chapter Eight
Protein

Protein as a complex of polypeptide

Polypeptides are often cross-linked by disulfide bonds between cysteine residues. And several polypeptides assemble to form a complex by hydrophobic and ionic interactions between hydrophobic and ionic side chains, respectively. Furthermore, polypeptides also interact with one another by hydrogen bonds. As a result, polypeptides form a complex, whose molecular mass sometimes becomes more than millions. We call the polypeptide complex a protein.

The role of protein

A protein plays a role in various life phenomena by using its complex molecular structure. For example, digestive peptic juices contains various digestive enzymes, which are protein complexes secreted in a stomach and an intestine to hydrolyze food materials such as polysaccharides, lipids, and proteins. A hair and wool are made from a protein called keratin. And, muscle is a treasure house of proteins, whose major components, actin and myosin, produces power for contraction. To tell the truth, almost all the life activities depend on the function of proteins. Proteins involve in regulation of life activities such as respiration, body temperature, growth, nutrition, learning, metabolism, excretion, motion, reproduction, and sleep.

Protein in life

Modern life science has discovered various life activities as a molecular mechanism of physiological materials. Especially, molecular biology succeeds to reveal the biochemical nature of physiological and genetic materials as proteins and nucleic acids, respectively, during the past seventy years. Proteins perform a daily life activity, and nucleic acid or genes, which design primary structure of proteins, are inherited from parents to children. The living things produce its healthy body from environmental physical materials such as water, air, and foods all through the life according to the plan of genes, which are inherited to the next generation.

Chapter Nine
Nucleic acids

Condensation of nucleotides

A nucleotide is a general name for a carbon compound with low molecular mass at around 300, which consists of a sugar, triphosphate and a base. There are two types of nucleotides, a ribonucleotide and a deoxyribonucleotide, which contain ribose and deoxyribose as a sugar, respectively. Also, there are four and three types of nucleotides according to a type of a base and the number of phosphates, respectively. Among them, di- and tri-phosphate nucleotides are able to build up polynucleotides or nucleic acids. Ribonucleotides and deoxyribonucleotides condense to produce ribonucleic acids (RNA) and deoxyribonucleic acids (DNA), respectively. Since DNA is concentrated in a nucleus of a eukaryotic cell, it is called nucleic acids.

A hydroxyl group at 3'-position of a ribose condensed with a triphosphate at 5'-position of another ribose to produce a phosphate di-ester bonding between the nucleotides releasing a pyrophosphate. This condensation is an exothermic reaction using a high energy of phosphate bonds, while it seldom occurs in water solution at room temperature. In a living cell, however, it does occur by a catalyst or a polymerase as discussed in detail later. Prokaryotes have a closed circular DNA made of millions of nucleotides, and eukaryotes have some linear DNA made of billions of nucleotides.

Structure of nucleic acids

Nucleic acids are synthesized by condensation of nucleotides, and a polynucleotide has triphosphate and hydroxyl groups at 5'- and 3'-position of ribose on each end, respectively. Messenger RNA and transfer RNA are single strand ribonucleic acids, which play a leading role in protein synthesis at ribosomes in a living cell. On the other hand, DNA or a genetic material consists of antiparallel double strands. DNA is the most important nitrogenous organic compound for life to be created during the material evolution in the solar system. Here, let us study the structure of DNA in detail.

DNA consists of antiparallel double strands of a polymer of deoxyribonucleotides. Both strands make a complementary pair by hydrogen bonds between bases, which are called base pairs. There are two types of base pairs, an adenine-thymidine pair and a guanine-cytosine pair. The former and latter contain two and three hydrogen bonds, respectively. It should be noted that a hydrogen transfer closed chain reaction occurs in any two hydrogen bonds within a base pair. In other words, a base is an enzyme of hydrogen transfer reaction for its complementary base to each other. As is the case of tubular β-sheets of polypeptides, the hydrogen transfer closed chain reaction causes a reverse chain reaction to stabilize the base pair structure in a single plane and consequently make hydrogen bonds rigid. Therefore, the double-strand DNA induces electromagnetic wave along its long axis to form a stand wave in it. Since electromagnetic induction make the base pairs diamagnetic to each other, each base pair tends not to overlap to one another along DNA axis and takes a distance between base pairs as long as possible. Consequently, double-stranded DNA tends to twist in right- or left-hand manner in a solid state and tends to elongate as long as possible in solution.

Replication of nucleic acids

It is obvious from the molecular structure of DNA that the complementary strand plays a role in catalyst for replication of DNA molecule. In order to synthesize a complementary DNA, a nucleotide is located on a matrix DNA to form a base pair, and condensed with a complementary polynucleotide, which had been synthesized and double stranded with a matrix DNA, at the 3'-end by DNA polymerase. A single strand matrix DNA works as a catalyst by putting nucleotides on the strand to form base pairs according to the nucleotide sequence. DNA polymerase stimulates the condensation reaction as catalyst, and matrix DNA also catalyzes the reaction by facilitating the accessibility of nucleotides on the polymerization site in a sequence specific manner. DNA is a catalyst for production of DNA itself. Furthermore, because DNA could reproduce its digital information or a nucleotide sequence, nucleic acids could perform its genetic function as a gene with nucleotide sequence information. The mechanism of reproduction of nucleic acids is well documented in several technical books, and readers may refer to them for further study.

Chapter Ten
Gene

Genetic information

After the discovery of nucleic acids as a genetic material by intracellular behavior and molecular structure, many scientists put all their energy into decoding the genetic information coded in the nucleotide sequence. Since nucleic acids themselves do not participate in physiological activities directly, it was a mystery how genes affect phenotypes of living things. Here in this chapter, let us see a summary of mechanism of genetic information and its application to life phenomena rather than a scientific history of molecular biology.

Genetic information is written in DNA as a base pair sequence, which is transcribed into RNA called messenger RNA or shortly mRNA. The nucleotide sequence of mRNA encodes an amino acid sequence of a protein. A protein is synthesized at ribosome and plays a various physiological functions in living things. Gene is a memory device encoding species-specific information of a nucleotide sequence, and a protein, which is produced from the genetic information, performs live activities. The secret of a seed is genetic materials or DNA, and flower blooms and produces seeds by functions of proteins.

Now, let us see the way of encoding genetic information in a nucleotide sequence of DNA.

Operon

Based on the genetics, it had been speculated that genes has some loci, which correspond to each phenotype of living thins.   Upon the progress on scientific research, it was revealed that a single locus corresponds to a polypeptide. In other words, the information of a polypeptide is encoded on DNA as a locus.   This genetic unit of information for a polypeptide is named an operon.   An operon contains several parts of information such as a promoter, a transcription starting position, a translation starting position, a protein sequence encoding region, a translation terminal site, and a transcription terminal site.   An operon corresponds to a locus of genes, and produces mRNA, which encodes a nucleotide sequence corresponding to an amino acid sequence of a polypeptide.

Genetic code

Comparison of a nucleotide sequence of mRNA and an amino acid sequence of a polypeptide clearly reveals that a sequence of three nucleotides codes one amino acid.   At the end of transcription, there is a sequence of three nucleotides, which does not correspond to any amino acid. All the textbooks of molecular biology present a decoding table of genetic codes.   A protein is synthesized from mRNA under the rule of the codon table.

Theoretically, the rule of genetic code could be diverged from species to species.   It is quite possible for scientists to find out living things, whose genetic coding table is quite different from that of human beings somewhere on earth or in the cosmos. However, after decipherment of genetic codes among various species of living things on earth, scientists realized a very important and common aspect of genetic codes. That is, all the living things on earth share a single coding table except for a few codons in a few species. This amazing discovery of common aspects of genetic code among all species on earth obviously indicates that all the living things on earth are derived from a common ancestor. All the living things on the earth are relatives or members of a large family. To our surprise, a non-life complex of DNA and proteins such as bacteriophages and viruses also uses exactly the same codon table as that of living things on earth.

Genetic molecule

In living things, genes are included in a cell. When living things grow, genes are replicated to be double and segregated equally into two daughter cells, which are divided by cytokinesis. If genes are segregated unequally or DNA is chemically modified, the daughter cells perform physiological phenotype different from parent' s one. In short, DNA is a minimum genetic molecule that could not be divided without changing its chemical properties.

It is far more difficult to show chemical properties of a single molecule than considered generally. Usually, scientists could show chemical properties of substances as a mass of uniform molecules. Actually, DNA is the very first to be assigned strictly as a single molecule in history of science.

Genetic mutation

Upon replication of genes, each base pair is processes as a digital information of G, A, T, or C. Even if a mismatched base is incorporated into DNA for the worst, the wrong nucleotide is promptly eliminated by exonuclease and a correct nucleotide is incorporated by a proofreading system. In such a way, living things guarantee a high level accuracy of gene replication, while mutations occur at a rate of once a ten million. And, genetic recombination happens sometimes and two DNA strands are crossing over with each other. Once unregulated recombination happens, the genomic structure changes dramatically. Furthermore, an excess chromosome could get into one daughter cell resulting in trisomy. It is also known that a small DNA or a transposon is excised from genome and inserted at another site to change genomic structure.

During the digital replication of genes, these mutations occur in DNA, which is expected to bring a diversity of living things causing a motive force for evolution.

Gene evolution

It is speculated that living things evolve gradually during a long period on earth. Evolution starts from a simple monad to multicellular organisms, and vertebrates are developed from fish, to amphibian, reptilian, bird, marsupial, and finally mammalian. Human beings appear on earth just millions of years ago. Some scientists insist that gene amplification and mutation forward evolution of living things. Systematic living things are created from chaotic small molecules and further evolved to human beings purposively. However, it does not sound for some biologists especially who consider living things as natural beings. They do not agree with the purposive evolution model from chemical substances to human beings because natural phenomena are generally getting uniform and orderly things become disorder as time goes by. In order to overcome this inconsistency between living things and natural phenomena, some scientists believe that supernatural power created living things with complex structure of genes purposively. Also biologists, who consider that it takes more than millions of years to have actual proof of evolution, emphasize that the evolution theory is not science because it is unverifiable.

How were human beings created? Were human beings created by someone or naturally? Was the creation performed by chance or inevitably? Here, I would just mention my personal belief. That is, a gene is created as a self-reproducing catalyst. Living things proliferate because of its natural properties, that is, self-reproduction. It is true that genes appear to proliferate purposively to increase the number of replica, while it is actually reproducible by its nature. It is exactly the nature of life. When genes are getting more complex, gene replication becomes more difficult. However, if the complexity has some merits for genes to survive or proliferate, the genes will evolve to be complex naturally. On the other hand, usually complexity has some demerits for proliferation and thus such complex genes will be eliminated naturally. Living things appear to repeat the diversity, natural selection, and stabilization of ecological system in nature as an evolutional cycle.

Chapter Eleven
Membrane

Fatty acid

A fatty acid is a hydrocarbon with a carboxyl group. A fatty acid is a major component of a triglyceride or a glycerol ester. A fatty acid is also a major component of lipids, which form a biological membrane. In a fatty acid molecule, a carboxyl group and an alkyl group are connected by a covalent bond. Within a water solution, fatty acids assembled together by hydrophobic interaction among their alkyl groups with hydrophilic carboxyl groups outside. Consequently, fatty acids form a micelle with alkyl and carboxyl groups inside and outside, respectively. Fatty acids could also form a single molecule membrane on the surface of water with their alkyl groups up.

A lipid is made from two fatty acids and one glycerol, which are connected by ester bonds. One remaining hydroxyl group of the glycerol could bind to some hydrophilic residue such as phosphate and sugar to produce phospholipid and glycolipid, respectively. Lipids tend to assemble into a plane type of micelle, that is, a membrane. Since lipids have various types of hydrophilic groups, a lipid membrane possesses a variety of chemical properties on its surface.

Triglycerides are made from three fatty acids and one glycerol, which are connected by ester bonds.

Triglycerides are energy stocks in fat tissues, and contain higher calories per mole than that of sugars or amino acids. Since triglyceride does not have any hydrophilic residues in it, the assembly of triglycerides becomes a large oil droplet with a few micrometers in a diameter.

Lipid bilayer

Lipids assemble and form a bilayer membrane in water solution naturally. In the membrane, a hydrophilic residue and two alkyl groups of a lipid molecule located on the surface and inside of the membrane, respectively. Inside the lipid membrane, the alkyl groups in one of the two layers contact with each other at the middle of the bilayer. Sometimes lipid bilayer spreads more than square micrometers extent as a plane membrane. Hydrophilic nature of both surfaces of a lipid bilayer stabilizes the bilayer structure of membrane in water solution. The lipid membrane tends to seal to form a vesicle, which prepares an aqueous environment inside the sphere. The lipid bilayer membrane plays a role not only in chemical activity of lipids but also in physical separation of inside from outside to prepare an intra-vesicular microenvironment.

Lipid bilayers perform various characteristics depending on the species and composition of constituent lipids. Lipids are classified to phospholipids and glycolipids according to their hydrophilic residues. And length of a carbon chain of alkyl groups of lipid determines thickness and a melting point of the lipid bilayer. Furthermore, a degree of unsaturation of the alkyl groups determines membrane fluidity and melting point. Moreover, the type of hydrophilic residues affects chemical properties of membrane surface. Here we just mention about the types of hydrophilic residues, and each unique characteristic on the lipid bilayer should be referred to appropriate technical textbooks.

A major component of lipid bilayer is phospholipid. In a phospholipid, various kinds of residues could bind to a phosphatidic acid by a phosphor ester bond. For example, an amino acid or serine makes an ester bond with phosphatidic acid to produce a phosphatidylserine. Also, ethanolamine, choline, and inositol make an ester bond with phosphatidic acid to produce phosphatidyl ethanolamine, phosphatidyl choline, and phosphatidyl inositol, respectively. Biophysical and biochemical nature of a membrane could be changed dramatically by phospholipid composition in the lipid bilayer. As mentioned later, metabolic products of lipid play an important role in transmission of environmental signs from outside into a cell as signal transduction molecules.

Biological membrane

Living things are either a monad or a multicellular organism. Each cell is enclosed with a lipid bilayer called a biological membrane. Biological membrane is not a simple bilayer of phospholipids, but it contains various lipids and its metabolites such as cholesterol, sphingolipid, diacylglycerol, and prostaglandin as well as phospholipids and glycolipids. The biological membrane also contains proteins such as ion channels and hormone receptors. And some secreted proteins such as extracellular matrix associate with the outer layer of biological membrane. On the other hand, various scaffold proteins associate with inner layer of a biological membrane, and some region of the membrane accumulates specific lipids and proteins to form so-called a membrane raft. Although a pure lipid bilayer does not permeate hydrophilic compounds at all, a biological membrane permeate proteins as well as various low molecular mass hydrophilic compounds such as water, salts, amino acids, and fatty acids because it has various types of transport systems in it.

The biological membrane not only plays a role in preparation of microenvironment inside of a cell but also absorbs nutrients from and excretes wastes to outer environment. Furthermore, a biological membrane is the first and most sensitive sensor for a cell to feel outside condition. The biological membrane is indispensable for living things to protect a cell, to sense outside conditions, to maintain intracellular conditions, to transport biomaterials into and out of a cell, and to proliferate with cell division.

Membrane protein

A biological membrane contains a lot of membrane proteins. A lipid bilayer exposes hydrophilic residues on its surfaces of both sides and packs hydrophobic carbon chains inside. Consequently, a membrane protein exposes hydrophilic and hydrophobic residues outside and inside of the lipid bilayer, respectively. In this way, some proteins associate to outer or inner surface of a membrane by hydrophilic interactions, while others are transmembrane or intramembrane proteins by hydrophobic interactions. Membrane-associated proteins are classified into several categories that are extracellular matrix proteins, intracellular scaffolding proteins, and intracellular signaling proteins. Transmembrane proteins include a variety of proteins such as receptors, ion channels, transporters, kinases, metabolic enzymes, and proteases. Membrane proteins play an indispensable role in maintenance of life functions in collaboration with biological membrane lipids.

Hydrophobic polypeptide in a transmembrane protein passes through a lipid bilayer. To reveal the structure and function of the transmembrane peptide in a lipid bilayer is one of the leading edge of life science research. Inside of the lipid bilayer is quite hydrophobic atmosphere by carbon chains of fatty acids. In the transmembrane polypeptide, the main chain takes a α-helical structure and the hydrophobic side chain interacts with alkyl group of fatty acids outside the helix. Generally, the α-helical structure is stabilized by a hydrogen transfer chain reaction through the helix as discussed previously. The chain reaction should be closed between outside and inside of the membrane. In other words, hydrogen should be added to a carbonyl group of the peptide bonds on one side and removed from amine of the peptide bonds on the other side of a biological membrane. Upon the chain reaction, an electric current flows in the transmembrane polypeptide to stabilize its α-helix structure. It is quite likely that ligand binding to an extracellular site of the receptor induces a hydrogen transfer chain reaction in the polypeptide, and an electric current flows across the transmembrane domain to an intracellular effector site, whose hydrogen transfer reaction stimulates activity of its intracellular target protein. Now, scientists pay attention to α-helical structure as a mediator of an allosteric effect in a multifunctional protein.

Membrane metabolism

A biological membrane plays an important role in receiving information from its environment. Biochemical nature of a lipid membrane directly affects the activity of membrane enzymes and intracellular proteins. In order to understand life, it is inevitable to reveal the molecular mechanism of membrane metabolism and its effect on cellular functions.

It is well known that the outer and inner layers contain different compositions of phospholipids. If the composition of phospholipids is artificially changed, the cell could not grow because of the inhibition of cytokinesis. And degradation of an inositol phospholipid by phospholipase C, two messengers, diacylglycerol and inositol-3-phosphate, are released from lipids. Diacylglycerol stimulates cell growth, and inositol-3-phosphate induces calcium release from endoplasmic reticulum by activation of an inositol-3-phosphate receptor. Furthermore, metabolites of phospholipids, prostaglandin and platelet activating factor, induce physiological actions of neurons and platelets, respectively.

Biological membrane is a physical barrier between life and environment, and also has various physiological functions to play indispensable role in life activities together with genes and proteins.

Chapter Twelve
Cell

A minimum unit of life

In previous chapters, we study biochemistry of living things from atom to life materials. It starts from the periodic table of elements to high molecular mass compounds and biological membranes. However, they are just materials but not living things. Bacteriophages and viruses are highly sophisticated complex materials consisting of proteins and nucleic acids, but do not perform any life activities by themselves and thus they are not living things. What is a minimum unit for life? In modern biology, it is called a cell. A monad or a single cell organism is real life. A monad contains genomes and proteins, which are surrounded by a biological membrane. A monad grows and proliferates naturally under a nutritional condition. A monad produces proteins to perform physiological activities in response to environmental conditions, and replicates genome, which is transferred to descendants. .

A cell contains all the life materials mentioned in previous chapters. All the materials in a cell are not only enveloped by a biological membrane but actively function in a cell organically. Life is not only the materials but also the activity of the materials in harmony. It is a long way from atom to life. Atoms gather and assemble into molecular substances, which produce various high molecular mass materials. However, it appears to be a large gap between materials and life. It is necessary to know chemical nature of all the materials in living things, but it is not enough for creation of the life.

Cell types

Cells are classified by several criteria. Largely, cells are classified into two types, that is, prokaryote and eukaryote. The former contains a genome in a cell while the latter contains a nucleus, in which a genome is included. Also, cells are classified into two types, that is, a monad and a multicellular organism. Further, cells are classified into two types, that is, an aerobic cell with mitochondria and an anaerobic cell without mitochondria. And, cells are classified into an autotrophic cell with chlorophyll and a heterotrophic cell without chlorophyll.

Cells perform an amazingly diverse complexity from cell to cell. Some monad such as a chlamydomonas and a paramecium develops highly complex body carrying sophisticated functions. Dictyostelium is a monad and also forms a multicellular fruiting body under starvation. Amazingly various types of living things exist on the earth. They survived during natural selection of evolution and now stably proliferate in the present ecological system.

Differentiation of a cell

A cell performs its specific nature depending on the species. Also, a cell shows various types of life activities after cell differentiation. The most famous, dramatic, and simple differentiation is the bacterial induction of β-galactosidase gene expression by nutritional lactose. In the absence of lactose in environment, bacteria do not express β-galactosidase, which is necessary to digest lactose as a carbon source nutrient. Upon the addition of lactose as nutrient, lactose induces the expression of its digestive enzyme, that is, β-galactosidase in bacteria. While the genome sequence of bacteria is not affected, bacteria quickly and dramatically change its digestive system from glucose to lactose according to the nutritional condition. We call this change of cell type differentiation. Almost all the cell differentiations take place by regulation of gene expression in a cell. The expression pattern is regulated not only by the genome but also by environmental signals. This is a very clever way of response for a cell to adjust to and survive in various environmental conditions.

A multicellular organism is usually developed from a single fertilized egg, and thus all the cells usually contain exactly the same set of genome in an individual organism. The cells, however, perform amazingly divergent physiological activities after development by proliferation and differentiation. It is almost impossible to believe that all the cells have been derived from a single egg. For example, muscle, blood, brain, and internal organs are derived from a single fertilized egg. The cells should take a long way of proliferation, migration, and differentiation to perform their highly sophisticated functions for their specific cellular activities.

Phylogenetic evolution

A cell produces a multicellular organism by embryogenesis. A cell is also suggested to produce various types of species by phylogenetic evolution during a line of generations. It is no doubt that genetic mutations drive phylogenetic evolution. It is a routine work for modern biologists to produce a new function in a cell by genetic manipulations. However, it takes more than hundreds of million years to evolve a cell from the first monad to human beings. Biological experiments are now on-going to reveal a molecular mechanism underlying the evolution of species and maintenance of ecological system.

On the other hand, biologists have a lot of evidence for evolution of species in the past by archeological specimens and fossils. History of living things of each period on the earth is carved there. It is impossible to know all species of living things at all period of history. However, we have an enough evidence of living things at some periods of the earth. Taken all together, it is revealed that several big bangs of species occurred in the history on earth to evolve the living things from fish, amphibian, reptilian, bird, marsupials, and mammals. Among the history, some species such as dinosaurs became extinct suddenly.

We still have lots of mystery of life such as a natural driving force of evolution and supernatural power of purposive evolution, and future progress of biological research is necessary to address these issues.

Universality of life

As we study on the fundamental nature of life in this book, all the physicochemical nature should be applicable not only on earth but also everywhere in the cosmos. Since biological elements in the periodic table are restricted in relatively low numbers of atomic mass units, which may exist universally in any solar systems, it is no wonder that living things exist on a planet in a solar system other than ours. Since the most fundamental nature of life materials is the self-reproducing catalysts, it is possible that life is born somewhere in our galaxy as a catalysts even in a different manner than that on the earth. For example, DNA has other base pairs than ours, and proteins might consist of more or less than twenty amino acids. It is also possible that racemic amino acids could produce proteins. Codon usage might not be the same as of ours. Still, the self-reproducing catalysts may proliferate by its physicochemical nature. It is nonsense to consider that life is a miracle only on the earth since the fundamental nature of living things should be applicable universally to any solar systems not only in our galaxy but also in the cosmos.

Future of life

At the end, I would like to speculate the future vision of life. It is not only fun but also fruitful to imagine the future possibility of life on the earth based on the physicochemical principle of life even if it is a scientific fiction but not science. It takes only three million years since human beings appeared on the earth. Homo sapience just appeared three hundred thousand years ago. During the last ten million years, mammalian has evolved so much. Especially, the brain has dramatically grown bigger and bigger. Since brain is the tissue of intelligence, development of brain should have merits for survival. In order to proliferate and prosper on the earth, in order to prepare for starvation, water flow, and meteorites, and in order to destiny to space, it is necessary to take a path from life to intelligence.

I guess that living things will further evolve to get higher intelligence to go to the space world. In order to get high intelligence, brain will grow bigger and also the neural network system will evolve not only quantitatively but also qualitatively in the near future.

References

1. The Origin of Species, Charles Darwin (1859)
2. The Double Helix, James Watson and Francis Click (1953)
3. General Chemistry, Linus Pauling (1970)

Acknowledgements

Just after Muhammad was born in AD570, ancient China and Kyoto started to build a checkered city with twelve gates for prosperity of Sino centrism. Twelve hundred years later, the West became economically rich after the Industrial Revolution, and also had a strong military after the 19th century. Western science, which started from physicochemical fields, progresses now in life science and neuroscience.

I wrote this book as an introduction for students to modern life science, which was established in the latter half of the 20th century. Also, I put some hints for development of new fields in life science somewhere in the book. It is my great pleasure if some readers of this book start to explore in life science to find out something new in this field.

Everyday, new life is born and joins our world. Life is really amazing, and should be grown with love of God. It is miracle to be born as living things, and it is more than great to be born as human beings. Believe in.

Finally, I would like to express my special thanks to my family, Yoko, Kodai, Arisa, and Hiroto for supporting my life and writing this book.

At home in Kasukabe

Apr 29th, 2015

Hiroyuki Aizawa

www.ingramcontent.com/pod-product-compliance
Lightning Source LLC
Chambersburg PA
CBHW040832180526
45159CB00001B/154